Doing Science

Adventures in Physical Science

Process-Oriented Activities for Grades 4–6

by Margy Kuntz

Fearon Teacher Aids
a division of
David S. Lake Publishers
Belmont, California

Cover and interior illustrations by Bradley Dutsch

Entire contents copyright © 1987 by David S. Lake Publishers, 19 Davis Drive, Belmont, California 94002. Permission is hereby granted to reproduce designated materials in this book for noncommercial classroom and individual use.

ISBN 0-8224-2319-7

Printed in the United States of America
1. 9 8 7 6 5 4 3 2

Contents

Introduction iv

Teacher's Guide 1

Activity Worksheets 23

                    Changing Matter 24

                    Name That Change! 25

                    Salt or Sugar? 26

                    The Litmus Test 27

                    Forceful Fields 28

                    A Magnetic Attraction 29

                    How Strong Is It? 30

                    What Type of Circuit? 31

                    Conductors and Insulators 32

                    Seeing Your Voice 33

                    Playing with Pitch 34

                    Sequencing Sounds 35

                    Making Music 36

                    Turning Up the Volume 37

                    Upon Reflection 38

                    Looking Through Lenses 39

                    Is It Symmetrical? 40

                    Red + Blue + Green = ? 41

                    Disappearing Colors 43

                    Is Seeing Believing? 44

Introduction

Teachers greatly influence children's interests. They expose children to new and stimulating topics and help them organize their knowledge. The *Doing Science* series was created to help you encourage children to be curious, to ask questions, to experiment, to learn, and to organize and integrate knowledge. This book will help you teach the processes of science—processes that can be integrated into all parts of our lives.

About This Book

Each **activity page** covers a topic that will easily fit into your science curriculum. The activities help students develop one or more of the following process skills:

- Observing—using the senses to gather information about objects and events.
- Comparing—identifying common and distinguishing characteristics among items or events.
- Measuring—comparatively or quantitatively describing the length, area, volume, mass, or temperature of objects.
- Classifying/Grouping—organizing information into logical categories.
- Sequencing—arranging items or events according to a characteristic.
- Collecting data—collecting and recording information obtained through observation.
- Organizing data—organizing data in a logical way so the results can be interpreted.
- Drawing conclusions—using the skills of inferring, predicting, and/or interpreting.

The **Teacher's Guide** will give you ideas for using each worksheet, including the main science concept, the process emphasis, and the materials list for each activity. The Teacher's Guide pages also include Discovery Questions—questions designed to make your students think and to encourage discussion. These questions are a mixture of specific-answer and open-ended questions that can be used either before, during, or after an activity. So while your students are doing science, they are also learning to think like scientists.

TEACHER'S GUIDE

Changing Matter

Concept

Matter can be changed.

Process Emphasis

Observing and collecting data

Materials

For each student:
- Activity worksheet, page 24
- Pencil or pen

For each group of five students:
- Eight paper cups
- Water
- Eight different materials (Suggested materials are: a tea bag, a piece of steel wool, a spoonful of oil, a piece of soap, a seltzer tablet, a dried bean, a piece of tissue paper, a spoonful of salt)
- A spoon

Procedure

1. Explain that matter can change in many ways. Ask students for examples and discuss their suggestions. Point out that matter changes because of the conditions that surround it or because of the substances it contacts.

2. Divide the class into groups of five and hand out the materials. Lead students through the directions on the worksheet. Once they have completed the activity they should answer the questions at the bottom of the page. (You might ask them to suggest ways in which some of the materials can be changed back into their original forms.)

3. If you want, discuss the differences between a physical change (one that changes the appearance but not the composition of a substance) and a chemical change (one that changes the composition of a substance).

Discovery Questions

- Could the changes that you observed have happened if you hadn't added water?
- How could the changes be speeded up?
- How could the changes be slowed down?

Name That Change!

Concept

Matter undergoes two main types of change—physical change and chemical change.

Process Emphasis

Classifying

Materials

For each student:
- Activity worksheet, page 25
- Pen or pencil

For the class:
- Several pieces of paper
- Scissors
- Match
- Tweezers
- Metal pan

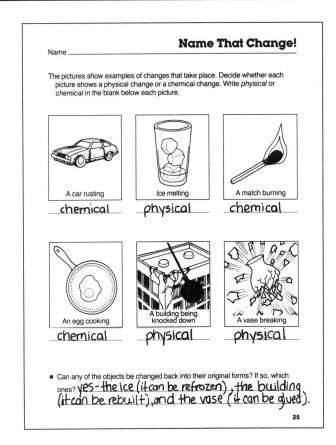

Procedure

1. Crumple a piece of paper into a ball. Ask students if the change you made resulted in a new material or if the material is the same as when you started. Now cut the paper into small pieces and repeat the question. Finally burn a small piece of the paper over the metal pan. Repeat the question. Explain that if you change the appearance of a substance (for instance, crumpling the paper or cutting it) but not the composition of the substance, you have made a change called a physical change. If you change the composition (for instance, burning the paper, which changes it into carbon dioxide and water) you have made a chemical change. Review other examples of the two changes.

2. Hand out the worksheet. Have students classify each change shown as a physical change or a chemical change.

Discovery Questions

- How could you physically change an egg?
- How could you chemically change an egg?

Salt or Sugar?

Concept

Matter has physical properties that can be observed.

Process Emphasis

Comparing and drawing conclusions

Materials

For each student:
- Activity worksheet, page 26
- Pen or pencil
- Labeled samples of sugar and salt
- Spoon
- Water
- Three paper cups
- Hand lens
- Unlabeled sample of *either* sugar or salt

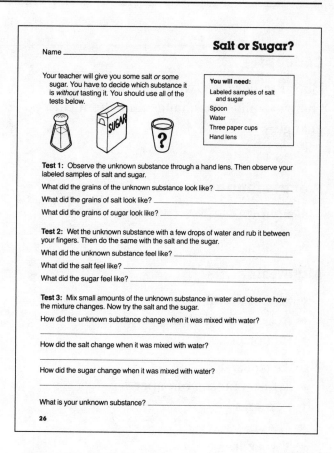

Procedure

1. Ask students to compare sugar and salt. (Most students will probably start by comparing the taste of the two substances.) Then ask them if they can think of any other ways besides taste to tell the difference between the two. Point out that both sugar and salt have properties that can be observed. (You may have to discuss the meaning of the word *property.*) Even though salt and sugar may look alike, if they are magnified they might look different. The feel of the two substances might be different, or the substances might behave differently when heated.

2. Hand out the materials. Remind students that they should never taste unknown substances, even if they think they know what the substances are. Lead students through the directions on the worksheet. When the class is finished, discuss the results.

Discovery Questions

- Why should you use more than one test to determine what a substance is?
- What are some other tests we could try?

The Litmus Test

Concept

Indicators can be used to identify acids, bases, or neutral solutions.

Process Emphasis

Collecting data, comparing, and drawing conclusions

Materials

For each student:
- Activity worksheet, page 27
- Pen or pencil

For each group of four students:
- Ten different samples of acids, bases, and neutrals
  Examples:
  Acids: lemon juice, aspirin-water, vinegar-water, pickle juice
  Bases: bleach-water, ammonia-water, soap-water, baking soda–water, milk of magnesia
  Neutrals: milk, plain water, neutralized mixture of an acid and a base (lemon juice mixed with baking soda)
- Ten paper cups
- Several strips of red and blue litmus paper

Procedure

1. Discuss acids, bases, and neutral solutions. Ask students if they know of any ways to find out if a solution is an acid or a base. Point out that students should never taste a substance to find out if it is an acid or a base. Explain that litmus paper contains a material called an indicator. It indicates, or shows, whether a substance is an acid or a base.

2. Hand out the materials. Tell students which substance is in each cup. Have them label the cups and write the names of the substances on their charts. Tell students that before they check each substance, they should guess (predict) whether the substance is an acid, a base, or a neutral solution. Have them record their guesses. When they have finished, they should test each solution using the litmus paper and record the answers.

Discovery Questions

- What are some ways in which acids are used?
- What are some ways in which bases are used?
- Why should you never taste a substance to determine whether it is an acid or a base?

Forceful Fields

Concept
Magnets have magnetic fields that surround them.

Process Emphasis
Collecting data and comparing

Materials
For each student:
- Activity worksheet, page 28
- Pen or pencil

For each pair of students:
- Two bar magnets
- Sheet of paper
- Iron filings (or finely snipped steel wool)
- Newspaper

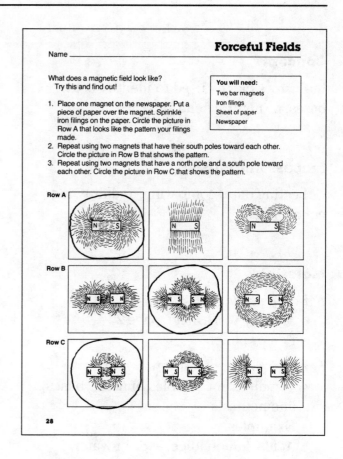

Procedure
1. Divide the class into pairs. Hand out the materials and lead students through the directions on the worksheet. (Students may have to try several times to get a clean picture of the magnets' fields.)

2. When students have completed the activity, discuss magnetic fields. Explain that the bar magnet attracts some iron filings and makes others line up around it. The region around the magnet that acts on the iron filings is the magnetic field.

Discovery Questions
- Do you think that a magnetic field around a horseshoe magnet would look the same as a field around a bar magnet?
- Do you think that the magnetic field of a stronger magnet would look the same as the magnetic field of a weaker magnet of the same shape?

A Magnetic Attraction

Concept

Magnets have different strengths.

Process Emphasis

Measuring and collecting data

Materials

For each student:
- Activity worksheet, page 29
- Pen or pencil

For each group of three students:
- Three different magnets labeled 1, 2, and 3
- Thirty small paper clips
- One large paper clip

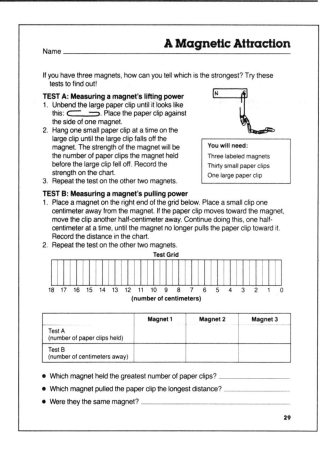

Procedure

1. Divide the class into groups of three. Hand each group the three magnets. Have them examine the magnets, and then ask them if they can tell you which magnet is the strongest. Discuss their answers.

2. Hand out the rest of the materials and the worksheet. Explain that they can measure the strength of each magnet using the materials. Guide students through the worksheet directions.

Discovery Questions

- What would happen to the strength of the magnets if we used heavier paper clips?
- What are some other units of measure we could use instead of paper clips?

How Strong Is It?

Concept

An electromagnet's strength can be changed by changing the length of the winding wire.

Process Emphasis

Measuring, collecting data, and organizing data

Materials

For each student:
- Activity worksheet, page 30
- Pen or pencil

For each group of two students:
- One "D" flashlight battery
- Five pieces of narrow gauge wire—one 2-foot length; one 3-foot length; one 4-foot length; one 5-foot length; one 6-foot length
- Forty small paper clips
- One large paper clip
- One iron nail or screw (1½-2½ inches long)
- Wire stripper

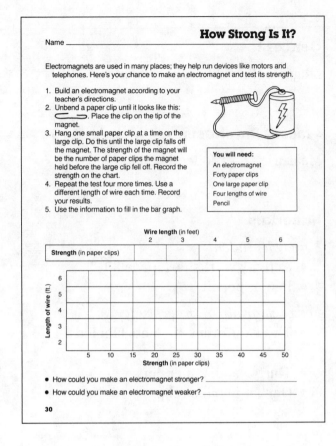

Procedure

1. Demonstrate how to build an electromagnet. Strip about 1 inch of insulation from each end of the wire pieces. Use the 2-foot piece of wire. Leave about 3 inches of wire free on both ends. Carefully wind the wire as tightly as possible around the screw or nail (see Figure A). When each end of the wire is placed against the end of the battery, the nail becomes a magnet. (Remind students not to keep the magnet on too long or the battery will wear out.)

2. Divide the class into groups of two and hand out the materials. Have each group make an electromagnet. Then guide students through the worksheet directions.

Discovery Questions

- Can you think of any other ways to make an electromagnet stronger? If so, what are they?
- How are magnets and electromagnets similar?
- How are magnets and electromagnets different?

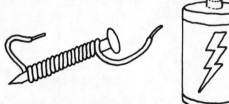

Figure A: An electromagnet

What Type of Circuit?

Concept

There are two types of electrical circuits—*series* circuits and *parallel* circuits.

Process Emphasis

Classifying

Materials

For each student:
- Activity worksheet, page 31
- Pen or pencil

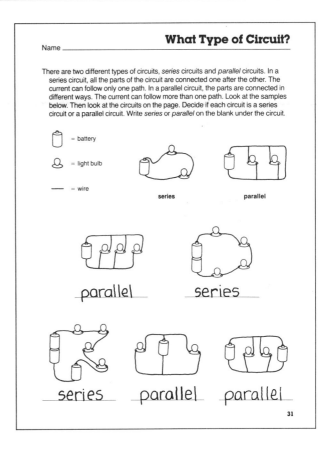

Procedure

1. Discuss the parts of an electrical circuit. Point out that a circuit usually has at least three parts—a source of electrical energy (batteries), a path of material for the current to travel in (wire), and something that uses the current (light bulb). Explain that there are two types of circuits—series circuits and parallel circuits. You might want to discuss the similarities and differences between the two circuits.

2. Hand out the worksheet. Have students classify each circuit as a series circuit or a parallel circuit.

Discovery Questions

- What happens if one part of the circuit is missing?
- Suppose you had a series circuit with one battery and two light bulbs. What would happen if one of the light bulbs burnt out?
- Suppose you had a parallel circuit with one battery and two light bulbs. What would happen if one of the light bulbs burnt out?

Conductors and Insulators

Concept

Some materials allow electricity to go through them. These are called *conductors*. Materials that don't allow electricity to go through are called *insulators*.

Process Emphasis

Collecting data and drawing conclusions

Materials

For each student:
- Activity worksheet, page 32
- Pen or pencil

For each group of four students:
- One "D" flashlight battery
- Light bulb and socket
- Insulated wire
- Electrical tape
- Aluminum foil
- Door key
- Paper cup
- Nickel
- Rubber band
- Pencil
- Wood or plastic ruler

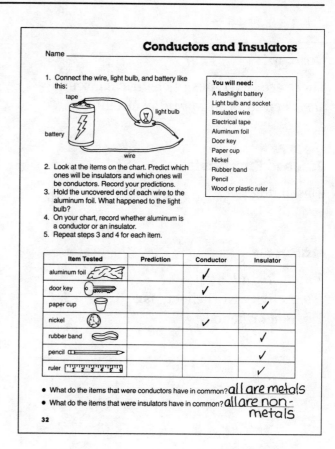

Procedure

1. Set up the circuit shown on the worksheet (page 32). Touch the two ends of the wire together. Explain that the electricity moves from one end of the battery to the other along the wire. The wire is called a conductor. It conducts electricity. Ask students if they can think of any other conductors. You should also discuss insulators.

2. Divide the class into groups of four and hand out the materials. Explain to students that they will be making a conductor tester. Guide students through the worksheet directions.

Discovery Questions

- What are some of the things we use insulators for?
- What are some of the things we use conductors for?

Seeing Your Voice

Concept
The pitch of a sound and the rate of vibration of the sound are directly related.

Process Emphasis
Observing, collecting data, and drawing conclusions

Materials
For each student:
- Activity worksheet, page 33
- Pen or pencil

For each group of two students:
- Coffee can or oatmeal container
- Can opener (if using coffee can)
- Scissors
- Large balloon
- Small piece of mirror (1–2 cm square)
- White glue
- Rubber band
- Flashlight (if it's a rainy day)

Procedure
1. Ask students if they can think of any ways to make sounds visible. Explain that one way to "see" the sound waves of your voice is to make a soundscope.

2. Divide the class into groups of two and hand out the materials. Have each group make a soundscope as described on page 33. Guide the students through the rest of the directions on the page.

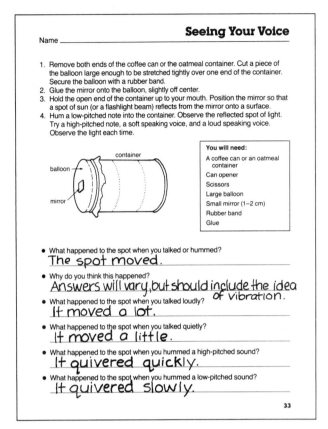

3. When the students have finished experimenting with their soundscopes, ask them if they can explain how the soundscope works. (The disturbance caused by the voice causes a stream of air to vibrate. The vibrating air hits the balloon and causes the balloon and the mirror to vibrate. The distance the light beam moves shows the loudness of the noise. The rate at which it quivers shows the pitch of the noise.)

Discovery Questions
- How is the balloon on the soundscope like an eardrum?

Playing with Pitch

Concept
Sounds differ in pitch.

Process Emphasis
Observing, comparing, and drawing conclusions

Materials

For each student:
- Activity worksheet, page 34
- Pen or pencil

For each group of four students:
- Three glasses (same size and shape)
- Water
- Measuring cup
- Three paper straws
- Scissors
- Three thicknesses of rubber bands
- Shoe box (without top)
- Spoon

Procedure

1. Divide the class into groups of four. Hand out the materials and lead students through the directions on the worksheet.

2. When students have finished the worksheet, discuss their results. Explain that the lowness or the highness of a sound is its *pitch*. The pitch of a sound is determined by how rapidly the source of the sound is vibrating. Point out that something that vibrates slowly produces a lower note and something that vibrates rapidly produces a higher note. You might want to tell students that the speed at which something vibrates is called its *frequency*.

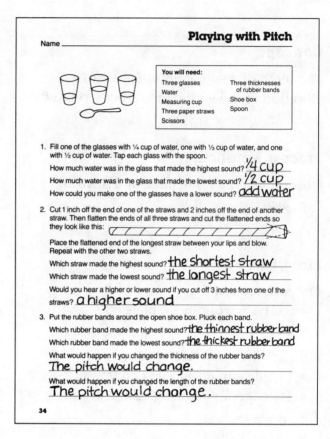

Discovery Questions
- What are the lowest and highest frequencies that people can hear?
- Why can cats and dogs hear sounds that we can't hear?

Sequencing Sounds

Concept

Sounds can be arranged according to pitch.

Process Emphasis

Sequencing

Materials

For each student:
- Activity worksheet, page 35
- Pen or pencil

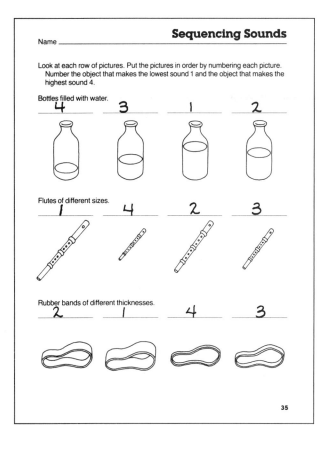

Procedure

1. Review the concept of pitch. Explain that something that vibrates slowly produces a lower note and something that vibrates quickly produces a higher note. You might want to discuss musical scales and how the scales are arranged by pitch.

2. Hand out the worksheet. Have students sequence the pictures by the pitch of the sounds the objects make. The lowest sound should be numbered 1 and the highest sound should be numbered 4.

Discovery Questions

- What are some things that make sounds with high pitches?
- What are some things that make sounds with low pitches?

Making Music

Concept

Musical instruments make sounds in several different ways.

Process Emphasis

Grouping

Materials

For each student:
- Activity worksheet, page 36
- Pen or pencil

For the class:
- Examples of a string, wind, and percussion instrument or pictures of these types of instruments

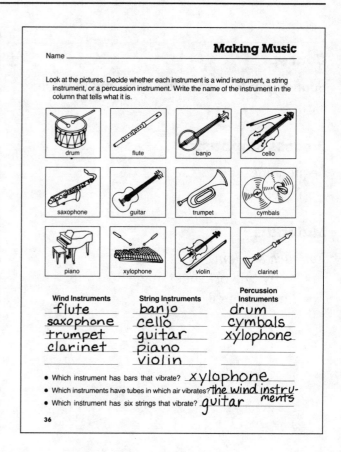

Procedure

1. Show students the three instruments or the pictures of the instruments. Ask them if they can tell you how the instruments make sounds. Discuss their answers and explain that the string instrument makes sounds because its strings vibrate, the wind instrument makes sounds because the column of air inside the tube vibrates, and the percussion instrument makes sounds because the object which is struck vibrates.

2. Hand out the worksheet. Have students write the name of each instrument in the appropriate category.

Discovery Questions

- What happens in all musical instruments that produce sound?
- How are voices like musical instruments?

Turning Up the Volume

Concept

The loudness of a sound can be measured in decibels.

Process Emphasis

Organizing data

Materials

For each student:
- Activity worksheet, page 37
- Pen or pencil

For the class:
- Piece of string

Procedure

1. Explain that when something vibrates and creates a sound, it moves back and forth past its original resting place. Demonstrate how this looks using the piece of string. Point out to students that the distance the string moves from its resting place is its *amplitude* (see Figure A). Loud sounds have big amplitudes and soft sounds have small amplitudes. Discuss the fact that scientists can measure the amplitude, or loudness, of a sound using a sound-level meter. The unit used to measure the loudness of sound is called the *decibel*.

2. Hand out the worksheet. Have students fill in the graph using the information on the page. When they have finished the graph they should answer the questions at the bottom of the page.

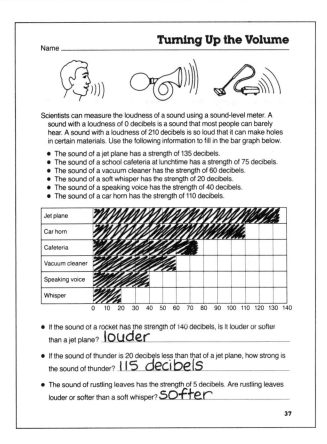

Discovery Questions

- What happens to your body when you hear loud sounds (over 70 decibels)?
- What is noise pollution?

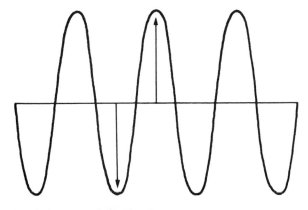

Figure A   **Amplitude**

Upon Reflection . . .

Concept

Light will reflect off a smooth surface to produce an image.

Process Emphasis

Classifying

Materials

For each student:
- Activity worksheet, page 38
- Pen or pencil

For the class:
- Aluminum foil

Procedure

1. Ask students to look around the room and identify objects that will reflect images. Explain that the images form when light bounces off a smooth surface. Discuss why some surfaces form better images than others. (When light rays hit smooth, shiny surfaces, all the rays reflect in the same direction. If the rays hit a rough surface, they reflect in many different directions.) Demonstrate using the piece of foil. First have students observe the smooth piece of foil. Then have them observe what happens to the image if the foil is crumpled.

2. Hand out the worksheet. Have students decide whether each object on the page reflects an image. They should write *image* or *no image* under each item.

Discovery Questions

- What kind of reflections do you see in a curved reflective surface (such as a teakettle or a hubcap)?
- Why is writing backward in a mirror?

Looking Through Lenses

Concept
Light bends when it goes through lenses.

Process Emphasis
Observing and drawing conclusions

Materials
For each student:
- Activity worksheet, page 39
- Three lenses—one flat, one concave, and one convex
- Five objects to look at (pencil, eraser, lettering on notebook, and so on)
- Pen or pencil

Procedure

1. Explain that a lens is a smooth piece of glass or plastic. When light goes through a lens, the light bends in special ways. Some lenses bend light so things look bigger. Some lenses bend light so things look smaller.

2. Hand out the materials. Have students follow the directions on the worksheet. When they have finished testing the lenses, they should answer the questions on the bottom of the page.

3. When the class is finished, you might want to explain how the lenses refract light.

Looking Through Lenses

Name _____

Your teacher will give you three types of lenses.
One will look like this: [ ] It is called a *flat* lens.
One will look like this:)[or][It is called a *concave* lens.
One will look like this: ([or () It is called a *convex* lens.

Look at five objects through each lens. Is the image of the object larger, smaller, or the same size as the object? Write down the results in the chart below.

	Flat Lens Larger, smaller, or same size?	Concave Lens Larger, smaller, or same size?	Convex Lens Larger, smaller, or same size?
Object 1	same size	smaller	larger
Object 2	same size	smaller	larger
Object 3	same size	smaller	larger
Object 4	same size	smaller	larger
Object 5	same size	smaller	larger

- What happens to the image of an object when you look at it through a flat lens? It stays the same size.
- What happens to the image of an object when you look at it through a concave lens? It becomes smaller.
- What happens to the image of an object when you look at it through a convex lens? It becomes larger.

Discovery Questions

- What type of lens do you think is in a microscope?
- What type of lens do you think is in a magnifying glass?
- What type of lens do you think is in a telescope?

Is It Symmetrical?

Concept

Mirrors can be used to determine symmetry.

Process Emphasis

Classifying

Materials

For each student:
- Activity worksheet, page 40
- Pen or pencil
- Pocket mirror

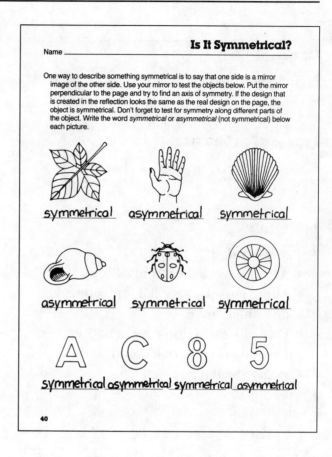

Procedure

1. Ask students to name things that mirrors can be used for (periscopes, looking at images, microscopes, and so on). Explain that mirrors can also be used to determine whether something is symmetrical or asymmetrical. Discuss symmetry. Point out that the word *symmetry* is used to describe something that has parts on opposite sides that look the same. You may want to draw some examples on the board.

2. Hand out the worksheet and mirrors. Have students experiment with each picture to see if they can find an axis of symmetry. (Some objects have more than one axis of symmetry.) If they can, they should write *symmetrical* on the blank under the picture. If the picture is not symmetrical, students should write *asymmetrical* on the blank.

Discovery Questions

- Are people symmetrical?
- What are some things you can find outside that are symmetrical?

Red + Blue + Green = ?

Concept

All the colors we see are made from a combination of three primary colors—red, blue, and green.*

Process Emphasis

Observing and collecting data

Materials

For each student:
- Activity worksheets, pages 41 and 42
- Red, blue, and green markers or crayons
- Scissors
- 8½" × 11" piece of cardboard
- Glue
- Four two-foot-long pieces of string
- Sharp pencil or nail

For the class:
- Prism

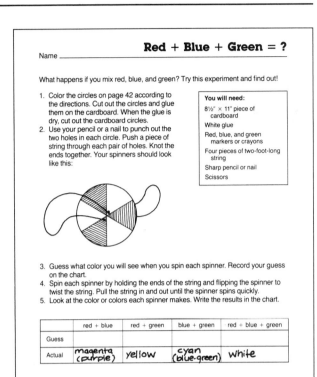

Procedure

1. Use the prism to introduce the visible spectrum. Explain that while we can distinguish between many different colors, our eyes are really only able to see three colors—red, blue, and green. From these three colors, our brains are able to "see" all the colors.

2. Hand out the materials. Lead the students through the directions on the worksheet. (Note: some of the colors produced by the spinners may look muddy—this is because the pigments in the markers are not very pure.)

3. When the students have finished the worksheet, ask them if they can explain how the color spinner works. (The spinner causes your eye to receive different color signals from the same place. They go by so quickly that your eye doesn't see them as flashes of different colors. Instead, it sees two or three signals from the same spot. Your brain combines these into one signal.)

Discovery Questions

- What happens when none of your color receptors receive signals?
- If you stare for a long time at a red spot and then look away, you'll see a green spot. What causes this?

*The primary colors of *light* are red, blue, and green—together these three colors create white light. The primary colors of *pigments* are red, blue, and yellow.

Disappearing Colors?

Concept

An object appears a certain color because it reflects that color and absorbs all the other colors.

Process Emphasis

Observing and collecting data

Materials

For each student:
- Activity worksheet, page 43
- Pen or pencil

For each group of four students:
- Flashlight
- Red, blue, and green cellophane
- Red, blue, green, and white construction paper
- Scissors
- Rubber band

Procedure

1. Ask students to make a list of all the colors they can see in the classroom. Darken the room. Ask students to make another list of all the colors that are now visible. Finally ask students what colors would be visible if the room were totally dark. Explain that we see the colors of objects because light is reflected from the object to our eyes. However, not all the light that reaches the objects is reflected. Discuss the primary colors of light (red, blue, and green) and how these colors can be mixed to form other colors.

2. Divide the class into groups of four and hand out the materials. Lead students

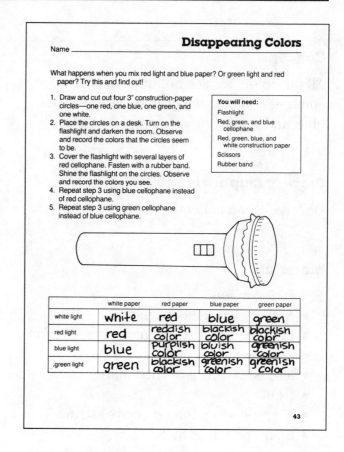

through the directions on the worksheet. (You should make the room as dark as possible. Also separate the groups of students so the light from one group doesn't interfere with the light from other groups.)

Discovery Questions

- What color would you see if you mixed red, blue, and green light?
- What color would you see if you mixed red and green light?
- What color would you see if you mixed blue and green light?
- What color would you see if you mixed red and blue light?

Is Seeing Believing?

Concept
Optical illusions occur when the brain misinterprets information from the eyes.

Process Emphasis
Observing and drawing conclusions

Materials
For each student:
- Activity worksheet, page 44
- Pen or pencil

Procedure

1. Hand out the worksheet. Have students answer all the questions on the page.

2. When the students are finished, tell them the answers. Explain that all the pictures on the page are called *optical illusions*. Ask students if they can guess why optical illusions fool us. Discuss their answers. Then explain that optical illusions fool our eyes and our brains. Sometimes our eyes are misled by slanted lines or arrows, so straight lines next to them look longer or shorter than they actually are (Illusions A and B on the worksheet). Sometimes we compare an object to the object right next to it, so a dot inside a large circle looks small while the same dot inside a small circle looks large (Illusion C). Some optical illusions occur because our brains get confused. We see more than one picture, but our brains aren't sure which picture they want to see (Illusion D).

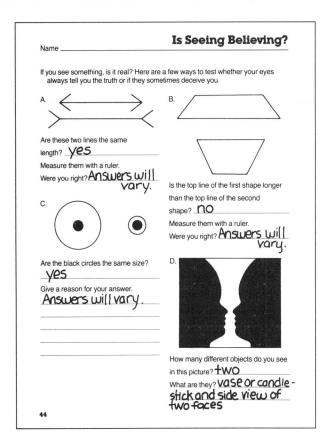

Discovery Questions
- What is the difference between an illusion and a mirage?
- Why does the moon look larger when it is just above the horizon than it does when it is higher in the sky?

ACTIVITY WORKSHEETS

Name _____

Changing Matter

1. Your teacher will give you eight different objects. Put one in each cup. Fill each cup half full of water.
2. Stir the contents of each cup and record what happens to the object.
3. Let your cups sit overnight. Observe the changes that occurred and record them.

You will need:
Eight paper cups
Water
Eight different materials
Pencil

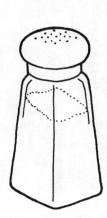

Material	What happened right after adding water	What happened after a day

- Can any of the materials be changed back into their original forms? If so, which ones? _____
- Are any of the changes permanent? If so, which ones? _____

Name That Change!

Name _____

The pictures show examples of changes that take place. Decide whether each picture shows a physical change or a chemical change. Write *physical* or *chemical* in the blank below each picture.

A car rusting
_____

Ice melting
_____

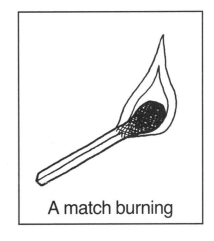
A match burning
_____

An egg cooking
_____

A building being knocked down
_____

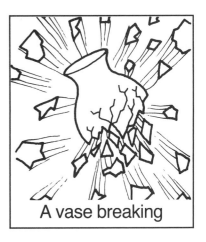
A vase breaking
_____

- Can any of the objects be changed back into their original forms? If so, which ones? _____
_____

Name _____

Salt or Sugar?

Your teacher will give you some salt *or* some sugar. You have to decide which substance it is *without* tasting it. You should use all of the tests below.

You will need:
Labeled samples of salt and sugar
Spoon
Water
Three paper cups
Hand lens

Test 1: Observe the unknown substance through a hand lens. Then observe your labeled samples of salt and sugar.

What did the grains of the unknown substance look like? _____

What did the grains of salt look like? _____

What did the grains of sugar look like? _____

Test 2: Wet the unknown substance with a few drops of water and rub it between your fingers. Then do the same with the salt and the sugar.

What did the unknown substance feel like? _____

What did the salt feel like? _____

What did the sugar feel like? _____

Test 3: Mix small amounts of the unknown substance in water and observe how the mixture changes. Now try the salt and the sugar.

How did the unknown substance change when it was mixed with water?

_____

How did the salt change when it was mixed with water?

_____

How did the sugar change when it was mixed with water?

_____

What is your unknown substance? _____

The Litmus Test

Name _____

What dissolves in water and forms an acid solution? A base solution? A neutral solution? Try this experiment and find out!

1. Your teacher will put a different substance in each cup. Write the name of the substance on the cup and under the *substance* column in the chart.
2. Guess whether each substance forms an acid, base, or neutral solution. Write your guess in the chart.
3. Use the litmus paper to test each solution. Record the color that the red and blue papers turn in the chart. Then use the Color Key to help you decide what type of solution each substance forms. Write the result in the chart.

You will need:

Ten substances dissolved in water
Ten paper cups
Red litmus paper
Blue litmus paper

Substance	Guess	Color of red litmus	Color of blue litmus	Type of solution

Color Key

The solution is **acid** if:	The solution is **base** if:	The solution is **neutral** if:
red litmus remains **red**	**red** litmus turns **blue**	**red** litmus remains **red**
blue litmus turns **red**	**blue** litmus remains **blue**	**blue** litmus remains **blue**

Forceful Fields

Name _____

What does a magnetic field look like?
Try this and find out!

You will need:
Two bar magnets
Iron filings
Sheet of paper
Newspaper

1. Place one magnet on the newspaper. Put a piece of paper over the magnet. Sprinkle iron filings on the paper. Circle the picture in Row A that looks like the pattern your filings made.
2. Repeat using two magnets that have their south poles toward each other. Circle the picture in Row B that shows the pattern.
3. Repeat using two magnets that have a north pole and a south pole toward each other. Circle the picture in Row C that shows the pattern.

Row A

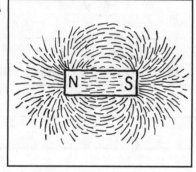 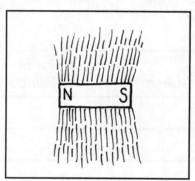 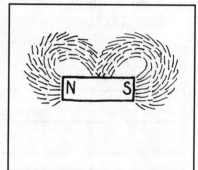

Row B

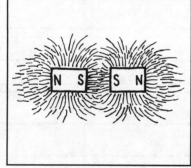 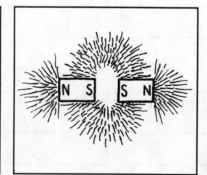 

Row C

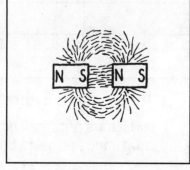  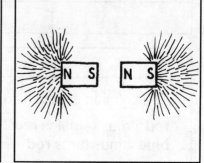

A Magnetic Attraction

Name _____

If you have three magnets, how can you tell which is the strongest? Try these tests to find out!

TEST A: Measuring a magnet's lifting power

1. Unbend the large paper clip until it looks like this: ⊂⊃. Place the paper clip against the side of one magnet.
2. Hang one small paper clip at a time on the large clip until the large clip falls off the magnet. The strength of the magnet will be the number of paper clips the magnet held before the large clip fell off. Record the strength on the chart.
3. Repeat the test on the other two magnets.

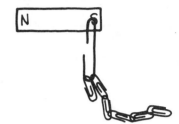

You will need:
Three labeled magnets
Thirty small paper clips
One large paper clip

TEST B: Measuring a magnet's pulling power

1. Place a magnet on the right end of the grid below. Place a small clip one centimeter away from the magnet. If the paper clip moves toward the magnet, move the clip another half-centimeter away. Continue doing this, one half-centimeter at a time, until the magnet no longer pulls the paper clip toward it. Record the distance in the chart.
2. Repeat the test on the other two magnets.

Test Grid

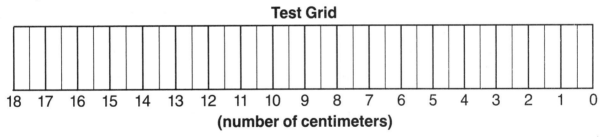

(number of centimeters)

	Magnet 1	Magnet 2	Magnet 3
Test A (number of paper clips held)			
Test B (number of centimeters away)			

- Which magnet held the greatest number of paper clips? _____
- Which magnet pulled the paper clip the longest distance? _____
- Were they the same magnet? _____

How Strong Is It?

Name _____

Electromagnets are used in many places; they help run devices like motors and telephones. Here's your chance to make an electromagnet and test its strength.

1. Build an electromagnet according to your teacher's directions.
2. Unbend a paper clip until it looks like this: ⌒⌐. Place the clip on the tip of the magnet.
3. Hang one small paper clip at a time on the large clip. Do this until the large clip falls off the magnet. The strength of the magnet will be the number of paper clips the magnet held before the large clip fell off. Record the strength on the chart.
4. Repeat the test four more times. Use a different length of wire each time. Record your results.
5. Use the information to fill in the bar graph.

You will need:
An electromagnet
Forty paper clips
One large paper clip
Four lengths of wire
Pencil

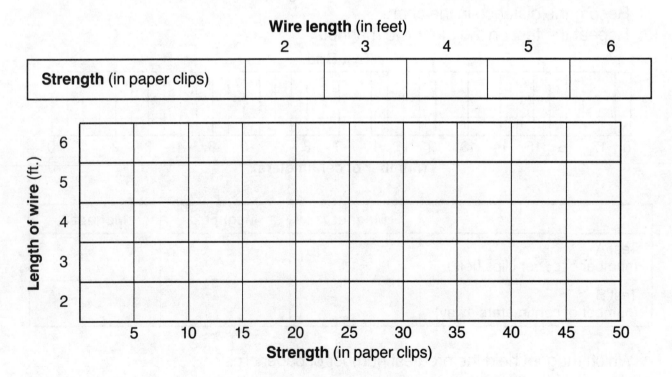

- How could you make an electromagnet stronger? _____
- How could you make an electromagnet weaker? _____

Name _____

What Type of Circuit?

There are two different types of circuits, *series* circuits and *parallel* circuits. In a series circuit, all the parts of the circuit are connected one after the other. The current can follow only one path. In a parallel circuit, the parts are connected in different ways. The current can follow more than one path. Look at the samples below. Then look at the circuits on the page. Decide if each circuit is a series circuit or a parallel circuit. Write *series* or *parallel* on the blank under the circuit.

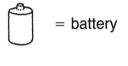 = battery

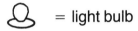 = light bulb

 = wire

series

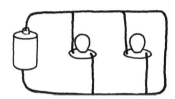
parallel

_____

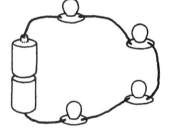
_____

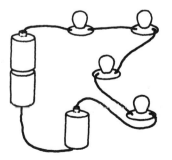
_____

_____

_____

Conductors and Insulators

Name _____

1. Connect the wire, light bulb, and battery like this:

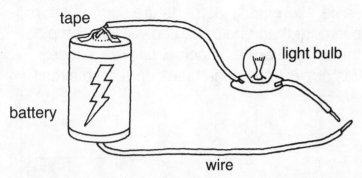

You will need:
A flashlight battery
Light bulb and socket
Insulated wire
Electrical tape
Aluminum foil
Door key
Paper cup
Nickel
Rubber band
Pencil
Wood or plastic ruler

2. Look at the items on the chart. Predict which ones will be insulators and which ones will be conductors. Record your predictions.
3. Hold the uncovered end of each wire to the aluminum foil. What happened to the light bulb?
4. On your chart, record whether aluminum is a conductor or an insulator.
5. Repeat steps 3 and 4 for each item.

Item Tested	Prediction	Conductor	Insulator
aluminum foil			
door key			
paper cup			
nickel			
rubber band			
pencil			
ruler			

- What do the items that were conductors have in common? _____
- What do the items that were insulators have in common? _____

32

Seeing Your Voice

Name _____

1. Remove both ends of the coffee can or the oatmeal container. Cut a piece of the balloon large enough to be stretched tightly over one end of the container. Secure the balloon with a rubber band.
2. Glue the mirror onto the balloon, slightly off center.
3. Hold the open end of the container up to your mouth. Position the mirror so that a spot of sun (or a flashlight beam) reflects from the mirror onto a surface.
4. Hum a low-pitched note into the container. Observe the reflected spot of light. Try a high-pitched note, a soft speaking voice, and a loud speaking voice. Observe the light each time.

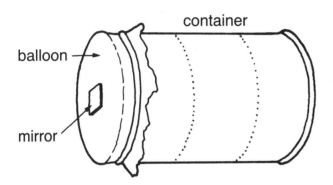

You will need:

A coffee can or an oatmeal container
Can opener
Scissors
Large balloon
Small mirror (1–2 cm)
Rubber band
Glue

- What happened to the spot when you talked or hummed?
  _____

- Why do you think this happened?
  _____

- What happened to the spot when you talked loudly?
  _____

- What happened to the spot when you talked quietly?
  _____

- What happened to the spot when you hummed a high-pitched sound?
  _____

- What happened to the spot when you hummed a low-pitched sound?
  _____

Playing with Pitch

Name _____

You will need:
- Three glasses
- Water
- Measuring cup
- Three paper straws
- Scissors
- Three thicknesses of rubber bands
- Shoe box
- Spoon

1. Fill one of the glasses with ¼ cup of water, one with ⅓ cup of water, and one with ½ cup of water. Tap each glass with the spoon.

   How much water was in the glass that made the highest sound? _____

   How much water was in the glass that made the lowest sound? _____

   How could you make one of the glasses have a lower sound? _____

2. Cut 1 inch off the end of one of the straws and 2 inches off the end of another straw. Then flatten the ends of all three straws and cut the flattened ends so they look like this:

   Place the flattened end of the longest straw between your lips and blow. Repeat with the other two straws.

   Which straw made the highest sound? _____

   Which straw made the lowest sound? _____

   Would you hear a higher or lower sound if you cut off 3 inches from one of the straws? _____

3. Put the rubber bands around the open shoe box. Pluck each band.

   Which rubber band made the highest sound? _____

   Which rubber band made the lowest sound? _____

   What would happen if you changed the thickness of the rubber bands? _____

   What would happen if you changed the length of the rubber bands? _____

Sequencing Sounds

Name _____

Look at each row of pictures. Put the pictures in order by numbering each picture. Number the object that makes the lowest sound 1 and the object that makes the highest sound 4.

Bottles filled with water.

   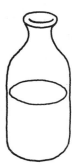

Flutes of different sizes.

Rubber bands of different thicknesses.

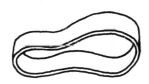 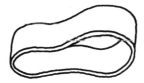  

Name _____

Making Music

Look at the pictures. Decide whether each instrument is a wind instrument, a string instrument, or a percussion instrument. Write the name of the instrument in the column that tells what it is.

drum

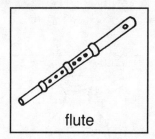
flute

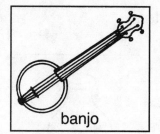
banjo

cello

saxophone

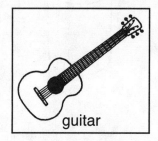
guitar

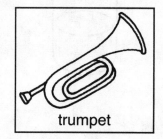
trumpet

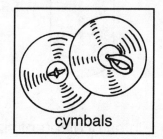
cymbals

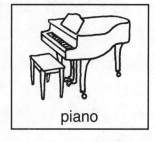
piano

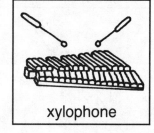
xylophone

violin

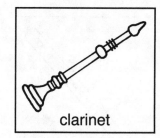
clarinet

Wind Instruments　　**String Instruments**　　**Percussion Instruments**

_____　　_____　　_____

_____　　_____　　_____

_____　　_____　　_____

_____　　_____　　_____

_____　　_____　　_____

- Which instrument has bars that vibrate? _____
- Which instruments have tubes in which air vibrates? _____
- Which instrument has six strings that vibrate? _____

Turning Up the Volume

Name _____

Scientists can measure the loudness of a sound using a sound-level meter. A sound with a loudness of 0 decibels is a sound that most people can barely hear. A sound with a loudness of 210 decibels is so loud that it can make holes in certain materials. Use the following information to fill in the bar graph below.

- The sound of a jet plane has a strength of 135 decibels.
- The sound of a school cafeteria at lunchtime has a strength of 75 decibels.
- The sound of a vacuum cleaner has the strength of 60 decibels.
- The sound of a soft whisper has the strength of 20 decibels.
- The sound of a speaking voice has the strength of 40 decibels.
- The sound of a car horn has the strength of 110 decibels.

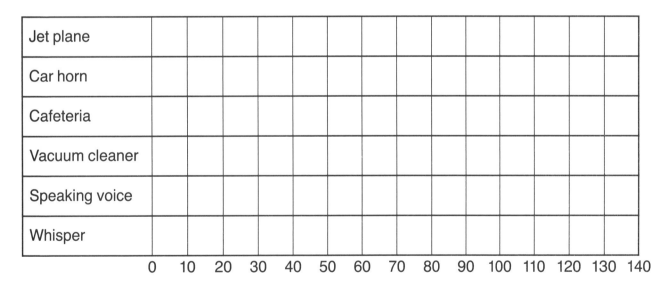

- If the sound of a rocket has the strength of 140 decibels, is it louder or softer than a jet plane? _____

- If the sound of thunder is 20 decibels less than that of a jet plane, how strong is the sound of thunder? _____

- The sound of rustling leaves has the strength of 5 decibels. Are rustling leaves louder or softer than a soft whisper? _____

Name _____

Upon Reflection . . .

Look at each object on the page. Decide if the object reflects an image. If it does, write *image* under the object. If it doesn't, write *no image* under the object.

  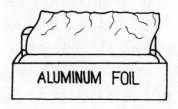

_____ _____ _____

_____ _____ _____

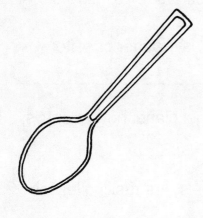 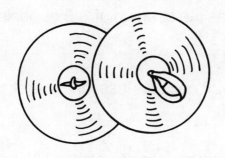 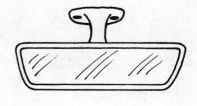

_____ _____ _____

Looking Through Lenses

Name _____

Your teacher will give you three types of lenses.

One will look like this: [] It is called a *flat* lens.

One will look like this:)[or][It is called a *concave* lens.

One will look like this: (] or () It is called a *convex* lens.

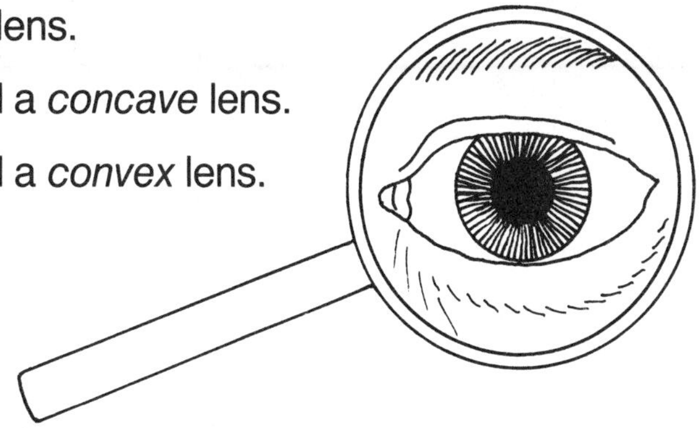

Look at five objects through each lens. Is the image of the object larger, smaller, or the same size as the object? Write down the results in the chart below.

	Flat Lens Larger, smaller, or same size?	**Concave Lens** Larger, smaller, or same size?	**Convex Lens** Larger, smaller, or same size?
Object 1			
Object 2			
Object 3			
Object 4			
Object 5			

- What happens to the image of an object when you look at it through a flat lens? _____

- What happens to the image of an object when you look at it through a concave lens? _____

- What happens to the image of an object when you look at it through a convex lens? _____

Name _____

Is It Symmetrical?

One way to describe something symmetrical is to say that one side is a mirror image of the other side. Use your mirror to test the objects below. Put the mirror perpendicular to the page and try to find an axis of symmetry. If the design that is created in the reflection looks the same as the real design on the page, the object is symmetrical. Don't forget to test for symmetry along different parts of the object. Write the word *symmetrical* or *asymmetrical* (not symmetrical) below each picture.

   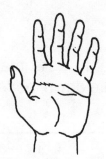   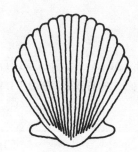

_____   _____   _____

      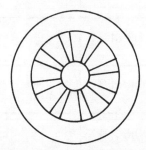

_____   _____   _____

_____   _____   _____   _____

Name _____

Red + Blue + Green = ?

What happens if you mix red, blue, and green? Try this experiment and find out!

1. Color the circles on page 42 according to the directions. Cut out the circles and glue them on the cardboard. When the glue is dry, cut out the cardboard circles.
2. Use your pencil or a nail to punch out the two holes in each circle. Push a piece of string through each pair of holes. Knot the ends together. Your spinners should look like this:

You will need:

8½" × 11" piece of cardboard

White glue

Red, blue, and green markers or crayons

Four pieces of two-foot-long string

Sharp pencil or nail

Scissors

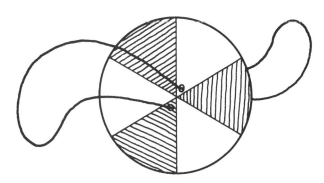

3. Guess what color you will see when you spin each spinner. Record your guess on the chart.
4. Spin each spinner by holding the ends of the string and flipping the spinner to twist the string. Pull the string in and out until the spinner spins quickly.
5. Look at the color or colors each spinner makes. Write the results in the chart.

	red + blue	red + green	blue + green	red + blue + green
Guess				
Actual				

Name _____

Red + Blue + Green = ?

Use these circles with page 41.
Color the circles as indicated.
Cut out the circles.

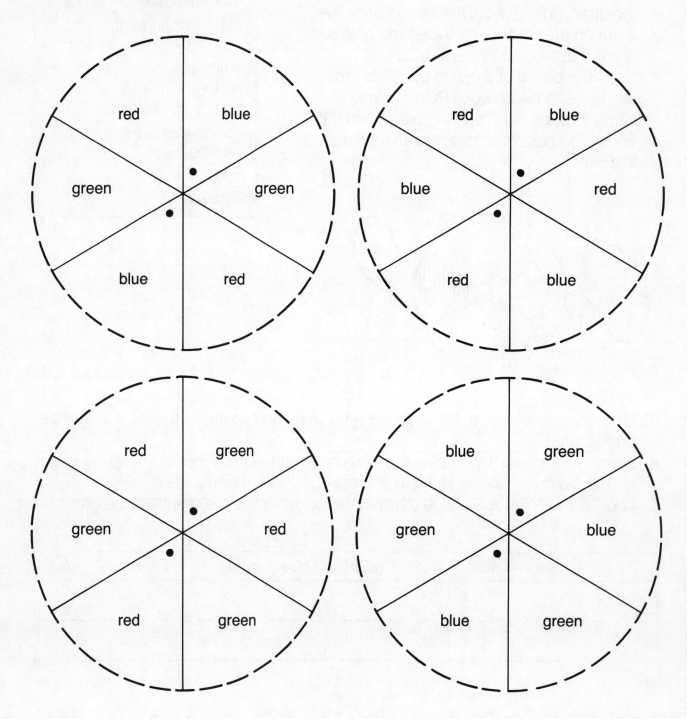

42

Adventures in Physical Science, © 1987 David S. Lake Publishers

Disappearing Colors

Name _____

What happens when you mix red light and blue paper? Or green light and red paper? Try this and find out!

1. Draw and cut out four 3" construction-paper circles—one red, one blue, one green, and one white.
2. Place the circles on a desk. Turn on the flashlight and darken the room. Observe and record the colors that the circles seem to be.
3. Cover the flashlight with several layers of red cellophane. Fasten with a rubber band. Shine the flashlight on the circles. Observe and record the colors you see.
4. Repeat step 3 using blue cellophane instead of red cellophane.
5. Repeat step 3 using green cellophane instead of blue cellophane.

You will need:
Flashlight
Red, green, and blue cellophane
Red, green, blue, and white construction paper
Scissors
Rubber band

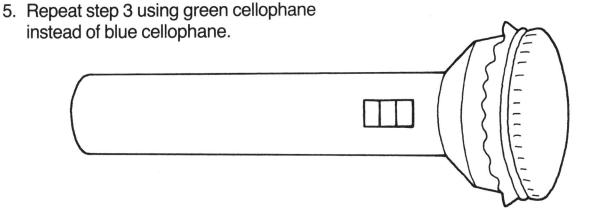

	white paper	red paper	blue paper	green paper
white light				
red light				
blue light				
green light				

Name _____

Is Seeing Believing?

If you see something, is it real? Here are a few ways to test whether your eyes always tell you the truth or if they sometimes deceive you.

A.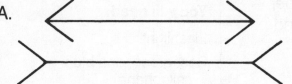

Are these two lines the same length? _____

Measure them with a ruler.

Were you right? _____

B.

Is the top line of the first shape longer than the top line of the second shape? _____

Measure them with a ruler.

Were you right? _____

C.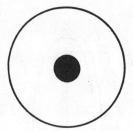

Are the black circles the same size? _____

Give a reason for your answer.

_____
_____
_____
_____

D.

How many different objects do you see in this picture? _____

What are they? _____

_____

44